COLLINS

Student Support Materials for

AQA

AS CHEMISTRY

Module 2: **Foundation Physical and Inorganic Chemistry**

John Bentham
Colin Chambers
Graham Curtis

This booklet has been designed to support the AQA Chemistry AS specification. It contains some material which has been added in order to clarify the specification.
The examination will be limited to material set out in the specification document.

Published by HarperCollins*Publishers* Limited
77–85 Fulham Palace Road
Hammersmith
London
W6 8JB

| www.**Collins**Education.com |
| Online support for schools and colleges |

ISBN 0 00 327702 X

British Library Cataloguing in Publication Data
A catalogue record for this publication is available from the British Library
Writing team: John Bentham, Colin Chambers, Graham Curtis, Geoff Hallas, David Nicholls, Andrew Maczek

Cover designed by Chi Leung
Editorial, design and production by Gecko Limited, Cambridge
Printed and bound by Sctoprint, Haddington

The publisher wishes to thank the Assessment and Qualifications Alliance for permission to reproduce the examination questions.

You might also like to visit

| www.**fire**and**water**.com |
| The book lover's website |

Other useful texts

Full colour textbooks
Collins Advanced Modular Sciences: Chemistry AS
Collins Advanced Science: Chemistry

Student Support Booklets
AQA Chemistry: Atomic Structure, Bonding and Periodicity
AQA Chemistry: Introduction to Organic Chemistry

What books do I need to study this course?

You will probably use a range of resources during your course. Some will be produced by the centre where you are studying, some by a commercial publisher and others may be borrowed from libraries or study centres. Different resources have different uses – but remember, owning a book is not enough – it must be *used*.

What does this booklet cover?

This *Student Support Booklet* covers the content you need to know and understand to pass the module test for AQA Chemistry Module 2: Foundation Physical and Inorganic Chemistry. It is very concise and you will need to study it carefully to make sure you can remember all of the material.

How can I remember all this material?

Reading the booklet is an essential first step – but reading by itself is not a good way to get stuff into your memory. If you have bought the booklet and can write on it, you could try the following techniques to help you to memorise the material:

- underline or highlight the most important words in every paragraph
- underline or highlight scientific jargon – write a note of the meaning in the margin if you are unsure
- remember the number of items in a list – then you can tell if you have forgotten one when you try to remember it later
- tick sections when you are sure you know them – and then concentrate on the sections you do not yet know.

How can I check my progress?

The module test at the end is a useful check on your progress – you may want to wait until you have nearly completed the module and use it as a mock exam or try questions one by one as you progress. The answers show you how much you need to do to get the marks.

What if I get stuck?

A colour textbook such as *Collins Advanced Modular Sciences: Chemistry AS* provides more explanation than this booklet. It may help you to make progress if you get stuck.

Any other good advice?

- You will not learn well if you are tired or stressed. Set aside time for work (and play!) and try to stick to it.
- Don't leave everything until the last minute – whatever your friends may tell you it doesn't work.
- You are most effective if you work hard for shorter periods of time and then take a (short!) break. 30 minutes of work followed by a five or ten minute break is a useful pattern. Then get back to work.
- Some people work better in the morning, some in the evening. Find out which works better for you and do that whenever possible.
- Do not suffer in silence – ask friends and your teacher for help.
- Stay calm, enjoy it and ... good luck!

The examiner's notes are always useful – make sure you read them because they will help with your module test.

The main text gives a very concise explanation of the ideas in your course. You must study all of it – none is spare or not needed.

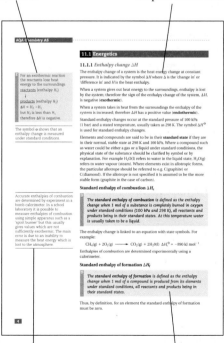

Further explanation references give a little extra detail, or direct you to other texts if you need more help or need to read around a topic.

There are rigorous definitions of the main terms used in your examination – memorise these exactly.

11.1 Energetics

11.1.1 *Enthalpy change ΔH*

The enthalpy change of a system is the heat energy change at constant pressure. It is indicated by the symbol ΔH where Δ is the 'change in' or 'difference in' and H is the heat enthalpy.

When a system gives out heat energy to the surroundings, enthalpy is lost by the system; therefore the sign of the enthalpy change of the system, ΔH, is negative (**exothermic**).

When a system takes in heat from the surroundings the enthalpy of the system is increased; therefore ΔH has a positive value (**endothermic**).

Standard enthalpy changes occur at the standard pressure of 100 kPa (1 bar) and a stated temperature, usually taken as 298 K. The symbol $\Delta H^{\ominus}$ is used for standard enthalpy changes.

The symbol $\ominus$ shows that an enthalpy change is measured under standard conditions.

Elements and compounds are said to be in their **standard state** if they are in their normal, stable state at 298 K and 100 kPa. Where a compound such as water could be either a gas or a liquid under standard conditions, the physical state of the substance should be clarified by symbol or by explanation. For example $H_2O(l)$ refers to water in the liquid state, $H_2O(g)$ refers to water vapour (steam). Where elements exist in allotropic forms, the particular allotrope should be referred to e.g. C(graphite) or C(diamond). If the allotrope is not specified it is assumed to be the more stable form (graphite in the case of carbon).

Standard enthalpy of combustion $\Delta H_c^{\ominus}$

Accurate enthalpies of combustion are determined by experiment in a bomb calorimeter. In a school laboratory it is possible to measure enthalpies of combustion using simple apparatus such as a 'spirit burner' but this usually gives values which are not sufficiently exothermic. The main error is due to an inability to measure the heat energy which is lost to the surroundings.

> **D**
>
> *The **standard enthalpy of combustion** is defined as the enthalpy change when 1 mol of a substance is completely burned in oxygen under standard conditions (100 kPa and 298 K), all reactants and products being in their standard states. At this temperature water is usually taken to be a liquid.*

The enthalpy change is linked to an equation with state symbols. For example:

$$CH_4(g) + 2O_2(g) \longrightarrow CO_2(g) + 2H_2O(l) \quad \Delta H_c^{\ominus} = -890 \text{ kJ mol}^{-1}$$

Enthalpies of combustion are determined experimentally using a calorimeter.

Standard enthalpy of formation $\Delta H_f^{\ominus}$

> **D**
>
> *The **standard enthalpy of formation** is defined as the enthalpy change when 1 mol of a compound is produced from its elements under standard conditions, all reactants and products being in their standard states.*

E For an element $\Delta H_f^{\ominus} = 0$.

Thus, by definition, for an element the standard enthalpy of formation must be zero.

The following is an example of a reaction for which the enthalpy change is the enthalpy of formation.

$$2Na(s) + C(graphite) + \frac{3}{2}O_2(g) \longrightarrow Na_2CO_3(s)$$

$$\Delta H_f^\ominus = -1131 \text{ kJ mol}^{-1}$$

Enthalpies of formation are usually determined indirectly using Hess's Law as explained in section 11.1.3 and can be found in data-book tables.

11.1.2 Calorimetry

The heat energy, q, required to change the temperature of a substance can be calculated using the formula:

$$q = m \times c \times \Delta T$$

where m is the mass of the substance in kg, c is the specific heat capacity in kJ K^{-1} kg^{-1} and ΔT is the change in temperature in K. For many chemical reactions in aqueous solution it can be assumed that the only substance heated is water, which has a specific heat capacity of 4.18 kJ K^{-1} kg^{-1}.

The heat energy, q, can be used to calculate an enthalpy change as shown in the two examples which follow.

Example 1

In an experiment 1.00 g of methanol (CH_3OH) was burned in air and the flame was used to heat 100 g of water, which rose in temperature by 42 K.

$$CH_3OH(l) + \frac{3}{2}O_2(g) \longrightarrow CO_2(g) + 2H_2O(g)$$

- Heat energy absorbed by the water $q = m \times c \times \Delta T$

$$= 0.1 \times 4.18 \times 42$$

$$= 17.6 \text{ kJ}$$

- Moles of methanol burned $= \dfrac{\text{mass}}{M_r}$

$$= \dfrac{1.00}{32}$$

$$= 0.031 \text{ mol}$$

- Enthalpy change $\Delta H = -\left(\dfrac{\text{heat energy}}{\text{moles of methanol}}\right)$

$$= -\dfrac{17.6}{0.031} \text{ kJ mol}^{-1}$$

$$= -570 \text{ kJ mol}^{-1}$$

For the purpose of this calculation, heat losses are ignored and the heat absorbed by the water container is regarded as negligible.

Note that the mass of water must be converted into kg (100 g = 0.10 kg). **E**

Note the **negative** sign because the reaction is exothermic.

Experiments like this are not usually very accurate. The answer has therefore been quoted to only two significant figures.

Example 2

In an insulated container, 50 cm^3 of 2.0 M HCl at 293 K were added to 50 cm^3 of 2.0 M NaOH also at 293 K. After reaction, the temperature of the mixture rose to 307 K.

$$HCl(aq) + NaOH(aq) \longrightarrow NaCl(aq) + H_2O(l)$$

- Temperature rise $\qquad\qquad\qquad\qquad \Delta T = 14 \text{ K}$

- Heat energy absorbed by the water $\quad q = m \times c \times \Delta T$

$$= 0.1 \times 4.18 \times 14$$

$$= 5.85 \text{ kJ}$$

- Moles of acid $\qquad\qquad\qquad\qquad$ = volume (in dm^3) $\times$ molarity

(= moles of alkali) $\qquad\qquad = \dfrac{50 \times 2.0}{1000}$

$$= 0.10 \text{ mol}$$

- Enthalpy change $\qquad\qquad\qquad \Delta H = -\left(\dfrac{\text{heat energy}}{\text{moles of acid}}\right)$

$$= -\dfrac{5.85}{0.10} \text{ kJ mol}^{-1}$$

$$= -59 \text{ kJ mol}^{-1}$$

> **E** The total volume of water in the reaction mixture is 100 cm^3. This has a mass of 0.1 kg. The water produced by the reaction is negligibly small. The heat capacity of the solution is assumed to be the same as that of water.

> **E** The enthalpy change is usually related to the 'moles of equation' as written. Again, this is an exothermic reaction so the sign of the enthalpy change is negative.

11.1.3 *Simple applications of Hess's Law*

The first law of thermodynamics is also known as the principle of conservation of energy.

> **D** **The first law of thermodynamics:**
> Energy can neither be created nor destroyed but it can be converted from one form into another.

Hess's Law, which is a special case of the first law, is stated below.

Hess's Law is a special case of the first law because it deals only with heat energy.

The overall enthalpy change for a multi-step reaction can be calculated using the expression $\Delta H = \Sigma(\Delta H_{(step\ 1)} + \Delta H_{(step\ 2)} + \ldots)$ where the symbol Σ means 'sum of'.

> **D** **Hess's Law**
> The enthalpy change of a reaction depends only on the initial and final states of the reaction and is independent of the route by which the reaction may occur.

It follows from Hess's Law that the enthalpy change of a reaction is the sum of the individual enthalpy changes for each step into which the reaction can be divided.

Hess's Law can be illustrated graphically as shown in Fig 1.

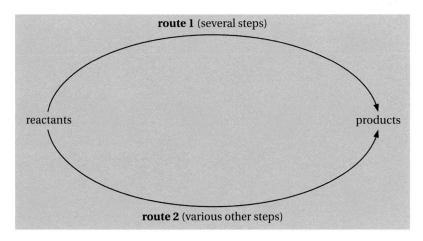

Fig 1
Hess's Law in terms of a heat cycle

$$\Delta H(\text{route 1}) = \Delta H(\text{route 2})$$

Hess's Law can be used to determine ΔH values for reactions where direct determination is difficult. For example, the enthalpy change for any reaction can be determined if the enthalpies of combustion of the reactants and the products are known. Thus the standard enthalpy of formation of methane can be calculated from standard enthalpies of combustion as follows.

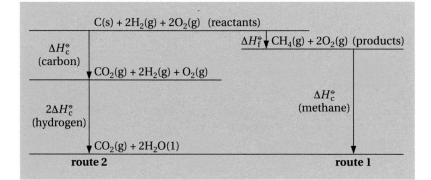

In the step where CH_4 is formed from $C(s)$ and $2H_2(g)$ the oxygen can be ignored because it is present in the reactants and products.

This cycle shows how an enthalpy change can be calculated from standard enthalpies of combustion. Note that in these diagrams, unlike the usual equations, the reactants are on one horizontal line and the products are on another.

$$\Sigma\Delta H(\text{steps in route 1}) = \Sigma\ \Delta H(\text{steps in route 2})$$

$$\therefore\ \Delta H_f^\circ(\text{products}) + H_c^\circ(\text{products}) = \Sigma\ \Delta H_c^\circ(\text{reactants})$$

$$\therefore\ \Delta H_f^\circ(\text{products}) = \Sigma\ \Delta H_c^\circ(\text{reactants}) - \Delta H_c^\circ(\text{products})$$

$$\therefore\ \Delta H_f^\circ(\text{methane}) = \Delta H_c^\circ(\text{carbon}) + 2 \times \Delta H_c^\circ(\text{hydrogen})$$

$$- \Delta H_c^\circ(\text{methane})$$

$$= -393 + (2 \times -285) - (-890)$$

$$= -73 \text{ kJ mol}^{-1}$$

Remember: $\Delta H_f(O_2)$ is zero because oxygen is an element.

In general, for any reaction $\mathbf{E}$
$\Delta H = \Sigma\ \Delta H_c(\text{reactants}) - \Sigma\ \Delta H_c(\text{products})$

The enthalpy change for reactions can also be determined from tabulated values of enthalpies of formation. For example, the enthalpy change for the reaction:

$$3CO(g) + Fe_2O_3(s) \longrightarrow 2Fe(s) + 3CO_2(g)$$

can be determined as follows.

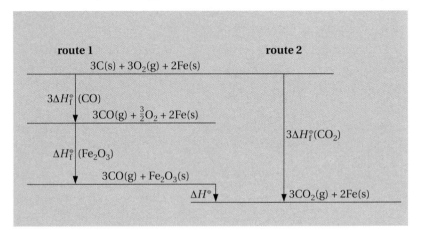

$$\Sigma \Delta H(\text{route 1}) = \Sigma \Delta H(\text{route 2})$$
$$\therefore \Delta H^\ominus + \Sigma \Delta H_f^\ominus(\text{reactants}) = \Sigma \Delta H_f^\ominus(\text{products})$$
$$\therefore \Delta H^\ominus = \Sigma \Delta H_f^\ominus(\text{products}) - \Sigma \Delta H_f^\ominus(\text{reactants})$$
$$\therefore \Delta H^\ominus = 3 \times \Delta H_f^\ominus(CO_2) - 3 \times \Delta H_f^\ominus(CO) + \Delta H_f^\ominus(Fe_2O_3)$$
$$= 3 \times -394 - (3 \times -111) - 822$$
$$= -27 \text{ kJ mol}^{-1}$$

11.1.4 *Bond enthalpies*

The bond enthalpy for a diatomic molecule is also known, more precisely, as the **bond dissociation enthalpy**. It refers to the enthalpy change for the following process where all species are in the gaseous state.

$$A—B(g) \longrightarrow A(g) + B(g) \qquad \Delta H = \text{bond enthalpy}$$

In polyatomic molecules it is convenient to use the term **mean bond enthalpy**.

Consider the following processes:

$$CH_4(g) \longrightarrow CH_3(g) + H(g) \qquad \Delta H = 423 \text{ kJ mol}^{-1}$$
$$CH_4(g) \longrightarrow C(g) + 4H(g) \qquad \Delta H = 1664 \text{ kJ mol}^{-1}$$

The second equation involves the breaking of all four carbon–hydrogen bonds. The mean bond enthalpy can therefore be determined by dividing the value of 1664 by 4. The bond dissociation enthalpy value (423 kJ mol^{-1}) is slightly different from the value for the mean bond enthalpy (416 kJ mol^{-1}). The first process involves the breaking of one C—H bond and the formation of a CH$_3$• radical. The second process does not involve the formation of hydrocarbon radicals; it leads to atomic species. The mean bond enthalpy is a useful quantity when calculating reaction enthalpy changes, but its use is only approximate since it is the average of values of bond dissociation enthalpies for a range of different compounds.

Mean bond enthalpies can be used to calculate the enthalpy change for simple reactions. The bond enthalpies of the reactant-bonds which are broken are added together. From that value is subtracted the sum of the bond enthalpies of the product-bonds which are formed, giving the overall enthalpy change. This is summarised in the following equation:

$$\Delta H = \Sigma(\text{enthalpy of bonds broken}) - \Sigma(\text{enthalpy of bonds formed})$$

For example, the enthalpy change for the following reaction can be calculated using the data from Table 1.

$$CH_4(g) + Cl_2(g) \longrightarrow CH_3Cl(g) + HCl(g)$$

$$\Delta H = \Sigma(\text{enthalpy of bonds broken}) - \Sigma(\text{enthalpy of bonds formed})$$

$$\Delta H = (C-H + Cl-Cl) - (C=Cl + H-Cl)$$

$$= (416 + 242) - (338 + 431)$$

$$= -111 \text{ kJ mol}^{-1}$$

Bond	C—H	C—Cl	Cl—Cl	H—Cl
Mean bond enthalpy/kJ mol^{-1}	416	338	242	431

E Bond enthalpy calculations apply only to reactions in the gaseous state.

E This calculation assumes that the C—H bond enthalpy in CH$_3$Cl is equal to that in CH$_4$; this is a good approximation. It is not always necessary to consider all the bonds in the reactants and products. In this example, the answer can be determined by considering only the bonds broken and those formed.

Table 1
Bond enthalpies

11.2 Kinetics

11.2.1 *Collision theory*

When two substances react, particles (molecules, atoms or ions) of one substance must collide with particles of the other. However, not all collisions result in a reaction, i.e. not all collisions are productive. This situation arises because particles will only react when they collide with sufficient energy. The minimum energy necessary for reaction is known as the **activation energy** (see also section 11.2.3).

11.2.2 *Maxwell–Boltzmann distribution*

In a sample of gas or liquid the molecules are in constant motion, and collide both with each other and with the walls of their container. Such collisions are said to be **elastic**, i.e. no energy is lost during the collision, but energy may be transferred from one molecule to another. Consequently, at a given temperature, molecules in a particular sample will have a spread of energies about the most probable energy. James Clark Maxwell and Ludwig Boltzmann derived a theory from which it is possible to draw curves showing how these energies are distributed. A plot of the number of molecules with a particular energy against that energy (see Fig 2) has become known as a Maxwell–Boltzmann distribution curve.

Fig 2
Distribution of energies at a particular temperature

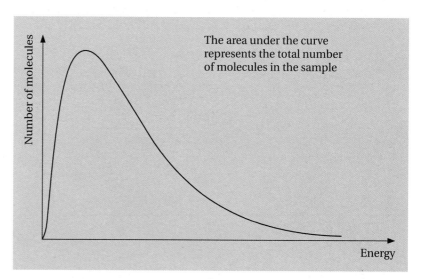

This distribution curve has several important features. There are no molecules with zero energy and only a few with very high energies. There is also no maximum energy for molecules – the curve in Fig 2 approaches zero asymptotically at high energy. The most probable energy of a molecule corresponds to the maximum of the curve as indicated in Fig 3.

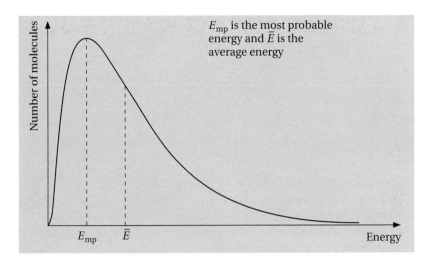

Fig 3
Most probable and average energies

E_{mp} is the most probable energy and $\bar{E}$ is the average energy

The effect of temperature variation on the Maxwell–Boltzmann curve

If the temperature of the sample is increased from T_1 to T_2, the average energy of the molecules increases, so that the most probable energy of the molecules increases. The spread of energies also increases and the shape of the distribution curve changes as shown in Fig 4. For a fixed sample of gas, the total number of molecules is unchanged so that the area under the curve remains constant (see also section 11.2.3).

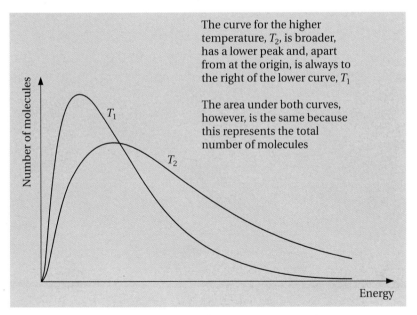

Fig 4
Distribution of energies at two temperatures

The curve for the higher temperature, T_2, is broader, has a lower peak and, apart from at the origin, is always to the right of the lower curve, T_1

The area under both curves, however, is the same because this represents the total number of molecules

11.2.3 *Factors affecting reaction rate*

The rate of a reaction can be defined as the change in concentration of a substance in unit time and has the SI units mol dm^{-3} s^{-1}.

When a graph is plotted of the concentration of a reagent or product against time, the rate of reaction at a particular time is given by the gradient of the graph at that time.

The rate is affected by:

- the concentration of reagents in solution or the pressure of gaseous reagents
- the surface area of any solid reagent
- the temperature
- the presence of a catalyst.

Collision theory can be used to explain how these factors affect the rate of reaction.

Concentration

Increasing the concentration of a reagent increases the number of particles in a given volume, and so increases the chance of productive collisions. This change increases the rate of reaction (whenever the reagent appears in the rate equation – see *Further Physical and Organic Chemistry*, section 13.1).

As a reaction proceeds, the reagents are used up, so their concentration falls. The rate is therefore at a maximum at the start of a reaction. On a concentration–time graph, the initial gradient is the most negative. The gradient falls to zero at the completion of the reaction, as shown in Fig 5.

Fig 5
Fall in concentration of reagent with time

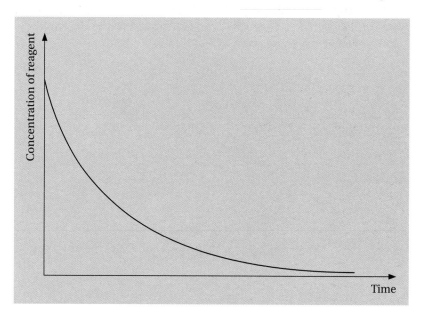

Surface area

When one reagent is a solid, the rate of its reaction with a gas, or with a substance in solution, is increased if the solid is broken into smaller pieces. This process increases the surface area of the solid and allows more collisions to occur with particles of the other reagent. For example, when a given mass of calcium carbonate is reacted with an excess of hydrochloric acid and the volume of carbon dioxide produced is plotted against time, Fig 6 shows that the gradient of the graph is much steeper when powdered carbonate is used (graph A) than when lumps are reacted (graph B). Note that the final volume of carbon dioxide evolved is the same in both

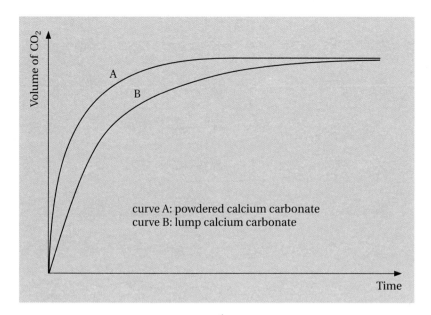

curve A: powdered calcium carbonate
curve B: lump calcium carbonate

Fig 6
Volume of CO_2 against time

experiments, since the same amount of calcium carbonate is used up in each case.

When an ionic solid is dissolved in a solvent, its particles are completely separated so that the rate is increased even further, and the reaction may become almost instantaneous. Precipitates form as soon as the correct solutions are mixed since the free ions in solution can easily collide and react.

Temperature

An increase in temperature always increases the rate of a reaction. According to the kinetic theory, the kinetic energy of the particles is proportional to the temperature. Particles have more energy at higher temperatures so they move about more quickly and there are more collisions in a given time.

More important, however, is the fact that particles will only react if, on collision, they have at least the minimum amount of energy, which is known as the activation energy.

> The **activation energy** of a reaction is the minimum energy required for the reaction to occur.

At higher temperatures, the mean energy of the particles is increased. The Maxwell–Boltzmann curves in Fig 7 overleaf show that, if the activation energy for a reaction is E_a, the number of molecules with energy greater than E_a (as shown by the shaded area) is much greater at temperature T_2 than at the lower temperature T_1.

Fig 7
Molecules with energy greater than
E_a at different temperatures

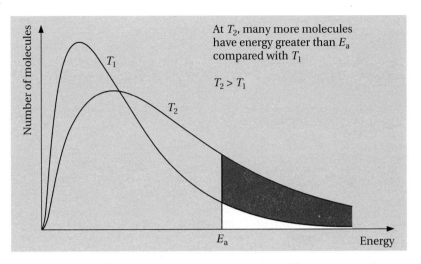

The number of collisions between molecules with sufficient energy to react, i.e. the number of productive collisions, and therefore the rate of reaction, is very much greater at the higher temperature. Consequently, small temperature increases can lead to large increases in rate, as shown in Fig 8.

Fig 8
Change of rate as temperature rises

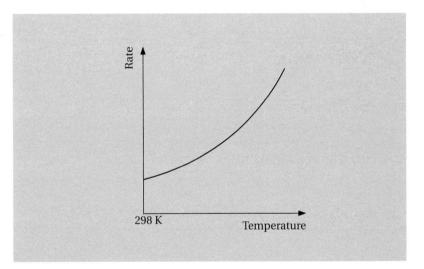

Many reactions – including several which are very exothermic – do not occur because the activation energy required is too high. For example, petrol reacts with oxygen in air in a very exothermic reaction, but a petrol–air mixture exists in the petrol tank of a car and will only react if sparked.

Catalysts
A catalyst is a substance which alters the rate of a reaction without itself being consumed during the reaction. Most of the catalysts used are positive catalysts: they increase the rate of reaction. An example of a negative catalyst, i.e. one which slows down a reaction, is antimony oxide which is used as a flame retardant in plastics such as PVC.

A positive catalyst operates by providing an alternative route or reaction mechanism which has a lower activation energy than the uncatalysed route. Fig 9 shows a reaction profile for a catalysed and an uncatalysed reaction. Note that the catalyst has no effect on the overall enthalpy change for the reaction. The catalyst also has no effect on the equilibrium position since this depends only on the relative energies of the reactants and products.

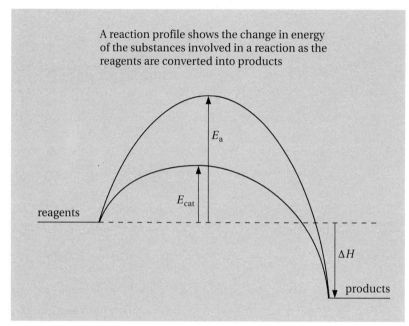

A reaction profile shows the change in energy of the substances involved in a reaction as the reagents are converted into products

Fig 9
Reaction profiles for catalysed and uncatalysed reactions

Catalytic action is covered in *Thermodynamics and Further Inorganic Chemistry*, section 14.4.6.

The reaction profile in Fig 9 shows a one-step catalysed reaction. In many cases the catalysed reaction occurs in more than one step and a double-humped reaction profile will be seen, as in Fig 10.

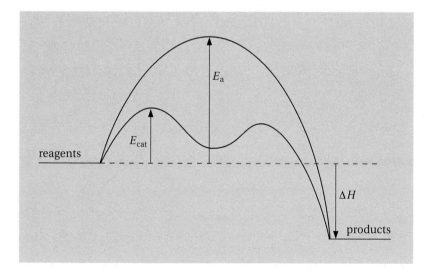

Fig 10
Reaction profiles

The Maxwell–Boltzmann curve in Fig 11 shows that a catalyst which lowers the activation energy from E_a to E_{cat} will produce many more molecules (the shaded area) that are able to react. In the presence of a catalyst, therefore, the rate is increased.

Fig 11
Effect of a catalyst on activation energies

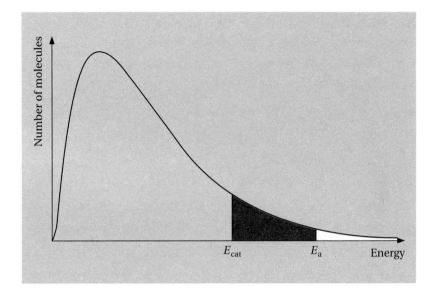

11.3 Equilibria

11.3.1 *The dynamic nature of equilibrium*

Many chemical reactions continue until one of the reactants is completely used up, and the reaction stops. Such reactions are said to *go to completion*. The reaction between magnesium and oxygen is a good example of a reaction which goes to completion.

$$2Mg + O_2 \longrightarrow 2MgO$$

Many other reactions, however, do not go to completion and are **reversible**. When the reactants and products have different colours, it is easy to demonstrate the reversibility of the reaction. For example, when dilute sulphuric acid is added to an aqueous solution containing yellow chromate(VI) ions, the following reaction occurs forming orange dichromate(VI) ions.

$$2CrO_4^{2-} + 2H^+ \longrightarrow Cr_2O_7^{2-} + H_2O$$

(yellow)   (orange)

If an aqueous solution of sodium hydroxide is now added to the orange solution, the reaction is reversed and yellow chromate(VI) ions are re-formed.

$$Cr_2O_7^{2-} + 2OH^- \longrightarrow 2CrO_4^{2-} + H_2O$$

(orange)   (yellow)

The overall reaction can be represented by the equation:

$$2CrO_4^{2-} + 2H^+ \rightleftharpoons Cr_2O_7^{2-} + H_2O$$

(yellow)   (orange)

> **E** Other reactions which can be used in the laboratory to illustrate the reversibility of reactions include $[Co(H_2O)_6]^{2+}$ with $Cl^-(aq)$ and $Fe^{3+}(aq)$ with SCN^-.

The $\rightleftharpoons$ sign is used to indicate that the reaction is reversible. By convention, the reaction shown as occurring from left to right in the equation is called the **forward reaction** whilst the reaction occurring in the opposite direction is called the **backward** or **reverse reaction**. Since the reaction still continues in both directions, it is said to be **dynamic**. When both reactions occur at the same rate, and the concentrations of the chromate(VI) and dichromate(VI) ions remain constant, a **chemical equilibrium** has been established. A chemical equilibrium can only be established if reagents are neither added to, nor taken from, the reaction mixture. If a reactant is added, the equilibrium position is then displaced.

When a chemical equilibrium is established:

- both reactants and products are present at all times
- the reaction is dynamic, i.e. it proceeds in both directions
- the concentrations of reactants and products remain constant.

If reactants are added or if products are removed, the equilibrium is displaced.

11.3.2 *The effects of changing reaction conditions*

The effect on the equilibrium position of the following will be considered:

- change in concentration
- change in pressure
- change in temperature
- addition of a catalyst.

For most reactions the qualitative effect of changing reaction conditions can be predicted using Le Chatelier's principle.

> **D**
> *Le Chatelier's principle: A system at equilibrium will react to oppose any change imposed upon it.*

The effect of a change in concentration

If, at a given fixed temperature, the concentration of any of the species involved in an equilibrium reaction is changed, then the concentrations of the other species must also change. Le Chatelier's principle can be used to deduce the changes which occur. For example, if the concentration of a reactant is increased, or the concentration of a product is decreased (e.g. by removing some of it), the position of equilibrium is displaced to the right and more product is obtained.

E This hydrolysis reaction can be studied in the laboratory. Weighed amounts of $CH_3COOC_2H_5(l)$ and water are mixed together in a conical flask, and an accurately measured volume of concentrated HCl is added as a catalyst (1.00 cm^3 is sufficient). The mixture is sealed with a bung and left overnight to reach equilibrium.

Titration with a standard solution of sodium hydroxide (1.00 M is ideal) gives the total acid present at equilibrium, both HCl and CH_3COOH. The volume of sodium hydroxide required by the HCl catalyst is also determined in a separate titration. Subtraction enables the volume of sodium hydroxide required to neutralise the acid produced, i.e. $CH_3COOH(l)$, to be determined. The number of moles of this acid can now be calculated and from this the number of moles of all the species present at equilibrium.

Consider the reaction:

$$CH_3COOC_2H_5(l) + H_2O(l) \rightleftharpoons CH_3COOH(l) + C_2H_5OH(l)$$

If a little more water or $CH_3COOC_2H_5(l)$ is added, some of the added reagent reacts and the equilibrium position is displaced to the right, so that the equilibrium yields of $CH_3COOH(l)$ and $C_2H_5OH(l)$ are increased. Similarly, if more $CH_3COOH(l)$ or more $C_2H_5OH(l)$ is added, the equilibrium is displaced to the left and the equilibrium mixture contains more $CH_3COOC_2H_5(l)$ and $H_2O(l)$.

N.B. If the reactants or products are gases, a change in the partial pressure of any gaseous species is equivalent to a change in the concentration of that species.

The effect of a change in total pressure

Changes in total pressure only have a significant effect on the composition of a mixture at equilibrium if the reaction involves gases. The changes observed are due to changes in the concentrations of the reactants. An increase in total pressure displaces the equilibrium in the direction of fewer moles of gas – the system responds by trying to decrease the pressure by reducing the number of moles of gas present. The converse of this statement also applies.

For example, consider the reaction:

$$CH_4(g) + H_2O(g) \rightleftharpoons 3H_2(g) + CO(g)$$

In this equation, the total number of moles of gaseous reactants is 2;

i.e. 1 mol of $CH_4(g)$ plus 1 mol of $H_2O(g)$, and the total number of moles of gaseous products is 4; i.e. 3 mol of $H_2(g)$ plus 1 mol of $CO(g)$.

Thus, at a given temperature, the equilibrium concentration of products can be increased by *reducing* the total pressure, so that the system responds by moving to the right to produce a greater number of moles of gas, in an attempt to increase the pressure.

The effect of a change in temperature
A change in temperature alters the rate of both the forward and the backward reactions. Since these are changed by different amounts, the endothermic position of the equilibrium is altered. Nevertheless, the simple rule which says that a system at equilibrium will react to oppose any change imposed upon it, can still be used to predict the effects of a change in temperature.

In an **exothermic forward reaction** heat energy is evolved. An increase in temperature is opposed by the reaction, which therefore moves in the endothermic direction with heat being absorbed. The equilibrium is displaced to the left and the equilibrium mixture contains a lower concentration of products. It is important to note, however, that although the new equilibrium mixture obtained at a higher temperature contains less product, the time taken to reach this new equilibrium is reduced because of the increased rate of the reaction.

N.B. The opposite is true if the temperature is *decreased*.

For example, consider the effect of a change in temperature on the exothermic reaction:

$$H_2(g) + I_2(g) \rightleftharpoons 2HI(g) \qquad \Delta H = -9.6 \text{ kJ mol}^{-1}$$

At 298 K, this equilibrium lies far to the right and the reaction mixture at equilibrium contains a high percentage of HI(g). At higher temperatures, the percentage of HI(g) present in the equilibrium mixture falls.

In an **endothermic forward reaction** heat energy is absorbed. An increase in temperature is opposed by the reaction, which therefore moves in the endothermic direction with heat energy being absorbed. In this case, the equilibrium is displaced to the right and the equilibrium mixture contains a higher concentration of products. Thus, for an endothermic reaction, an increase in temperature increases the rates of the forward and backward reactions unequally and also increases the equilibrium concentration of products.

N.B. The converse is true if the temperature is *decreased*.

Consider, for example, the endothermic reaction:

$$N_2(g) + O_2(g) \rightleftharpoons 2NO(g) \qquad \Delta H = 180 \text{ kJ mol}^{-1}$$

At 298 K, the equilibrium lies so far to the left that the equilibrium mixture contains almost no NO(g). Increasing the temperature to 1500 K does increase the equilibrium yield of NO(g) but this is still too small for the direct combination of nitrogen and oxygen to be an economically viable method of preparing NO(g).

If pressure is lowered, the rate of the reaction decreases.

Changes in pressure have no effect on the position of this equilibrium since the number of moles of gas on each side of the equation is the same.

19

The effects of changes in temperature on equilibria can be summarised as follows: an *increase* in temperature always displaces the equilibrium in the *endothermic* direction.

Table 2
Effects of temperature changes on equilibria

Reaction enthalpy	Change in temperature	Displacement of equilibrium	Yield of product	Rate of attainment of equilibrium
exothermic	increased	to the left	reduced	increased
exothermic	decreased	to the right	increased	reduced
endothermic	increased	to the right	increased	increased
endothermic	decreased	to the left	reduced	reduced

The effect of a catalyst

The addition of a catalyst to a mixture at equilibrium has no effect on the composition of the equilibrium mixture. This is because a catalyst increases the rate of both the forward and the backward reactions, which are equal at equilibrium. Hence the equilibrium position is achieved more quickly, but the composition of the equilibrium mixture is unchanged.

11.3.3 *The importance of equilibria in industrial processes*

Many chemicals are manufactured on a large scale. The processes used are designed to give the optimum yield. All the factors which affect the position of a particular equilibrium reaction must be carefully considered. The synthesis of ammonia from nitrogen and hydrogen, by the process originally discovered by Fritz Haber in 1908, can be used as an illustration.

Since ammonia can readily be oxidised to nitric acid, the discovery made by Haber enabled Germany to manufacture both fertilisers – essential for food production – and explosives on a large scale throughout World War I.

$$N_2(g) + 3H_2(g) \rightleftharpoons 2NH_3(g) \quad \Delta H = -92 \text{ kJ mol}^{-1}$$

The manufacturer aims for the highest possible yield, in the shortest possible time, for the lowest possible cost.

Operating pressure

The equation for the synthesis of ammonia shows that four moles of gaseous reactants (one of nitrogen and three of hydrogen) form two moles of gaseous products (ammonia). Le Chatelier's principle states that a system at equilibrium will react to oppose any change imposed upon it, therefore the synthesis of ammonia is favoured by high pressure. In practice, the pressure used is often around 2.0×10^4 kPa. Although higher pressures produce greater equilibrium yields of ammonia, the cost of generating such pressures is usually uneconomic; the pressure of 2.0×10^4 kPa is a good compromise between cost and equilibrium yield of ammonia.

Operating temperature

The ammonia synthesis reaction is an exothermic process ($\Delta H = -92$ kJ mol^{-1}). As the forward reaction gives out heat energy, the application of Le Chatelier's principle predicts correctly that this reaction will be opposed by an increase in temperature. Therefore the best equilibrium yield of ammonia is obtained at a low temperature.

However, at low temperatures the rate of reaction is low, and although a high equilibrium yield of ammonia can be achieved, it may take a long time to reach equilibrium. Increasing the temperature speeds up the rate of attainment of equilibrium, but reduces the equilibrium yield. A compromise is clearly necessary. The usual operating temperature is in the range 650–720 K.

Use of a catalyst

The rates of both the forward and backward reactions are increased to the same extent by the use of a catalyst; hence the time taken for the reaction to reach equilibrium is reduced. New catalysts – often a mixture of several compounds – are continually being developed, and their composition is often a closely guarded industrial secret. The catalyst usually employed for ammonia synthesis is based on iron, with potassium hydroxide added to promote its activity. When this catalyst is used, the usual operating temperature for the synthesis reaction is in the range 650–720 K.

Computer simulations of the Haber process enable the user to investigate how changes in pressure, temperature and catalyst will influence the yield of ammonia.

11.4 Redox reactions

11.4.1 *Oxidation and reduction*

The term **redox** is used for reactions which involve both **reduction** and **oxidation**. Originally the term oxidation was applied to the formation of a metal oxide when a metal reacted with oxygen. For example:

$$2Mg + O_2 \longrightarrow 2MgO$$

The reverse of this reaction was given the name reduction, and reducing agents were substances which removed oxygen. Hence, in the reaction:

$$Fe_2O_3 + 2Al \longrightarrow Al_2O_3 + 2Fe$$

aluminium behaves as a reducing agent, and in the reaction:

$$CuO + C \longrightarrow Cu + CO$$

carbon is the reducing agent.

The role of hydrogen as a reducing agent was recognised and the following definitions were given:

- oxidation – the addition of oxygen (or the removal of hydrogen)

- reduction – the removal of oxygen (or the addition of hydrogen).

Now consider the reactions:

$$SO_2 + H_2O + HgO \longrightarrow H_2SO_4 + Hg$$

$$SO_2 + 2H_2O + Cl_2 \longrightarrow H_2SO_4 + 2HCl$$

Clearly in both reactions there is an oxidation of $SO_2(aq)$ to $SO_3(aq)$, i.e. H_2SO_4, and in the second case the oxidising agent must be chlorine. The use of oxidation states enables the definitions to be improved.

11.4.2 *Oxidation states*

For a simple ion, the oxidation state is the charge on the ion:

Na^+, K^+, Ag^+	have an oxidation state of	$+1$
$Mg^{2+}, Ca^{2+}, Ba^{2+}$	have an oxidation state of	$+2$
F^-, Cl^-, I^-	have an oxidation state of	-1
O^{2-}, S^{2-}	have an oxidation state of	-2

> **E** Oxidation state can also be called oxidation number.

The oxidation state of the central atom in a complex ion (an ion consisting of several atoms) is the charge it would have if it were a simple ion and not bonded to other species. Table 3 gives some data which can be used to establish the oxidation state of an atom in a complex ion.

Table 3
Data for assignment of oxidation states

Species	Oxidation state
elements not combined with others	0
oxygen in compounds	-2
hydrogen in compounds, except in metal hydrides	$+1$
hydrogen in metal hydrides	-1
Group I metals in compounds	$+1$
Group II metals in compounds	$+2$

> **E** Combined oxygen in peroxides has an oxidation state of -1.

Calculation of the oxidation state of a combined element in an oxo-ion

Using the data given in Table 3, the oxidation state of an atom in a complex ion can be calculated. For example, the oxidation state of phosphorus in PO_4^{3-} is determined as follows.

The overall charge is -3, therefore:

oxidation state of phosphorus $+ (4 \times$ oxidation state of oxygen$) = -3$

Hence: oxidation state of phosphorus $- 8 = -3$

Thus: oxidation state of phosphorus $= +5$

Some more examples of oxo-ion complexes are given in Table 4. The sum of the oxidation states in a neutral compound is zero; the sum of the oxidation states in an ion is equal to the overall charge of the ion.

Species	Number of oxygen atoms	Total oxidation number due to oxygen	Overall charge on the ion	Oxidation state of central atom	Name of species
SO_4^{2-}	4	-8	-2	$+6$	sulphate
NO_3^-	3	-6	-1	$+5$	nitrate
ClO_3^-	3	-6	-1	$+5$	chlorate(V)
ClO^-	1	-2	-1	$+1$	chlorate(I)

Table 4
Calculation of the oxidation state of some central atoms in oxo-ions

The ending -ate means that the ion has a negative charge. **E**

More precisely, SO_4^{2-} should be called sulphate(VI) and NO_3^- should be called nitrate(V) but these oxidation states are omitted in common usage. **E**

11.4.3 *Redox equations*

In the reaction:

$$Fe_2O_3 + 2Al \longrightarrow Al_2O_3 + 2Fe$$

the changes which occur are shown in the half-equations below:

$$Fe^{3+} + 3e^- \longrightarrow Fe$$
$$Al \longrightarrow Al^{3+} + 3e^-$$

The oxidation state of iron changes from $+3$ in Fe_2O_3 to zero in the uncombined metal, i.e. a reduction occurs. The oxidation state of aluminium changes from zero in the uncombined metal to $+3$ in Al_2O_3 and the aluminium metal is oxidised.

The oxidation of $SO_2(aq)$ to $SO_3(aq)$ by chlorine can be understood by the use of oxidation states. In the reaction:

$$Cl_2 + SO_2 + 2H_2O \longrightarrow H_2SO_4 + 2HCl$$

chlorine is reduced from oxidation state zero to oxidation state -1, as shown by the half-equation below:

$$Cl_2 + 2e^- \longrightarrow 2Cl^-$$

The oxidation state of sulphur is increased from $+4$ to $+6$ by oxidation:

$$SO_2 + 2H_2O \longrightarrow H_2SO_4 + 2H^+ + 2e^-$$

This is the Thermite reaction, which can be demonstrated in the laboratory, but it is essential that full safety precautions are taken.

Redox reactions can be summarised as shown:

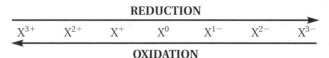

REDUCTION

$$X^{3+} \qquad X^{2+} \qquad X^{+} \qquad X^{0} \qquad X^{1-} \qquad X^{2-} \qquad X^{3-}$$

OXIDATION

Oxidation is the process of electron loss. Reduction is the process of electron gain.

Identifying redox reactions

Redox reactions can readily be understood by the use of the above definitions.

Example 1

Explain why the following is a redox reaction.

$$Mg + 2HCl \longrightarrow MgCl_2 + H_2$$

Answer

In this reaction, the oxidation state of hydrogen, combined in hydrochloric acid, changes from $+1$ to zero in the uncombined element, i.e. the hydrogen ion is reduced by the magnesium metal. The uncombined magnesium metal, in oxidation state zero, is changed into combined magnesium in magnesium chloride, with an oxidation state of $+2$. Thus the magnesium metal is oxidised by the hydrochloric acid.

The use of an ionic equation makes this redox reaction easy to recognise.

$$Mg + 2H^+ \longrightarrow Mg^{2+} + H_2$$

Example 2

Explain why the following is a redox reaction.

$$MnO_2 + 4HCl \longrightarrow MnCl_2 + Cl_2 + 2H_2O$$

Answer

Deductions using the oxidation states of combined oxygen and chlorine show that in this reaction manganese is reduced from oxidation state $+4$ to $+2$ by the chloride ions in hydrochloric acid. Some of the chlorine, combined in HCl, is converted into the uncombined element chlorine. This change, from oxidation state -1 to zero, is due to oxidation by manganese(IV) oxide. The equation below, written as a simple ionic equation, shows the changes in oxidation state.

$$Mn^{4+} + 2Cl^- \longrightarrow Mn^{2+} + Cl_2$$

Example 3

Explain why the following is not a redox reaction.

$$MgO + 2HCl \longrightarrow MgCl_2 + H_2O$$

Answer

At an initial glance this might appear to be a redox reaction as the magnesium oxide 'loses' oxygen. There is, however, no change in the oxidation state of any of the elements present. The reaction is that of an acid with a base, resulting in the formation of a salt and water.

$$MgO + 2H^+ \longrightarrow Mg^{2+} + H_2O$$

Half-equations for redox reactions

Earlier in this section the equations:

$$Fe^{3+} + 3e^- \longrightarrow Fe$$

$$Al \longrightarrow Al^{3+} + 3e^-$$

were given. These are examples of half-equations. The overall equation for a redox reaction can be separated into two half-equations; one shows reduction, the other oxidation. In each case the half-equation is balanced using electrons, so that the overall charge on both sides of the equation is the same.

These equations are often much simpler than molecular equations because they only show the actual species involved in the reaction. It is only necessary to know the initial and final species in a redox reaction to be able to construct half-equations for both processes.

The construction of half-equations for reactions

When constructing any half-equation, the following points must be observed:

- only *one* element in a half-equation changes oxidation state
- the half-equation must balance for atoms
- the half-equation must balance for charge.

When constructing a half-equation for reactions occurring in aqueous solution, water provides a source of oxygen and any 'surplus' oxygen is converted into water by reaction with hydrogen ions from an acid. Applying these rules, half-equations for the reduction of any species can be deduced.

Example

Deduce the half-equation for the reduction, in acid solution, of NO_3^- to NO.

Answer

The oxidation state of nitrogen changes from $+5$ in NO_3^- to $+2$ in NO and nitrogen is reduced. The oxidation state of the oxygen is still -2, but two of the three oxygens combine with four hydrogen ions (provided by the added acid) to form two water molecules.

$$NO_3^- + 4H^+ \longrightarrow NO + 2H_2O$$

This half-equation now balances for atoms but not for charge, with a total charge of $+3$ on the left hand side and zero on the right hand side. Three electrons must be added to the left hand side to give the balanced half-equation:

$$NO_3^- + 4H^+ + 3e^- \longrightarrow NO + 2H_2O$$

The construction of overall equations for redox reactions

The overall equation for any redox reaction can be obtained by adding together two half-equations, making sure that the number of electrons given by the reducing agent exactly balances the number of electrons accepted by the oxidising agent.

Example

When chlorine gas is bubbled through an aqueous solution of potassium bromide, the solution turns yellow as bromide ions are oxidised to bromine by chlorine, which is itself reduced to chloride ions. Write half-equations for the oxidation of bromide ions and for the reduction of chlorine, and use these to deduce an overall equation for the reaction.

Answer

The half-equation for the reduction of chlorine is:

$$Cl_2 + 2e^- \longrightarrow 2Cl^-$$

The half-equation for the oxidation of bromide ions to bromine is given below.

$$2Br^- \longrightarrow Br_2 + 2e^-$$

The number of electrons gained by chlorine in the first equation must equal the number of electrons given by two bromide ions in the second equation. The overall equation can therefore be obtained by simply adding together the two half-equations. The overall equation does not involve electrons, because they cancel out.

$$Cl_2 + 2e^- \longrightarrow 2Cl^-$$
$$2Br^- \longrightarrow Br_2 + 2e^-$$
$$\overline{Cl_2 + 2Br^- \longrightarrow 2Cl^- + Br_2}$$

E Potassium ions take no part in this reaction and can be omitted from the equations. They are called spectator ions.

Example

When concentrated nitric acid is added to copper metal, copper is oxidised to oxidation state +2 and nitric acid is reduced to nitrogen(IV) oxide. Write half-equations for the oxidation of copper and for the reduction of nitric acid, and use these to deduce an overall equation for the reaction.

Answer

The half-equation for copper is given below and shows that the uncombined copper metal is oxidised by loss of electrons.

$$Cu \longrightarrow Cu^{2+} + 2e^-$$

The reduction reaction involves the reduction of nitrogen from +5 in HNO_3 to +4 in NO_2.

$$HNO_3 + H^+ + e^- \longrightarrow NO_2 + H_2O$$

It is necessary to add H^+ ions to the left hand side of this half-equation to combine with the 'surplus' oxygen (see p. 25).

The two half-equations can now be combined to give an overall redox equation for the reaction. When this is done, the number of electrons required by the species being reduced and the number given by the species being oxidised must be the same, so that the equation for the overall redox reaction does not contain electrons.

In this example, the half-equation for the reduction of nitric acid must be doubled, so that the overall equation will correctly show that the two electrons lost by copper metal are accepted by nitric acid.

The overall equation for the reaction is given by addition:

$$Cu \longrightarrow Cu^{2+} + 2e^-$$
$$2HNO_3 + 2H^+ + 2e^- \longrightarrow 2NO_2 + 2H_2O$$
$$\overline{2HNO_3 + 2H^+ + Cu \longrightarrow Cu^{2+} + 2NO_2 + 2H_2O}$$

Other useful examples include the reduction of concentrated H_2SO_4 to either SO_2 or S or H_2S when warmed with solid NaI (see section 11.5.3).

In certain cases, where the same species appears on both sides of an overall equation obtained by the addition of two half-equations, it is necessary to cancel such species to give the final equation. Water molecules and hydrogen ions are the most common species which need to be treated in this way. Ions which take no part in the reaction are also omitted.

11.5 Group VII, the halogens

The halogens form a family of non-metallic elements which show clear similarities and well-defined trends in their properties as the relative atomic mass increases.

11.5.1 *Trends in physical properties*

It is not necessary to remember data, but it is necessary to know and be able to explain the observed trends. Some data are given in Table 5.

Table 5
Group VII data

Element	Atomic number	Outer electrons	Atomic radius/nm	Radius of X^- ion/ nm	Boiling point/K	Electro-negativity
F	9	$2s^2 2p^5$	0.071	0.133	85	4.0
Cl	17	$3s^2 3p^5$	0.099	0.180	238	3.0
Br	35	$4s^2 4p^5$	0.114	0.195	332	2.8
I	53	$5s^2 5p^5$	0.133	0.215	457	2.5

Trend in electronegativity of the halogens

The term electronegativity was introduced in *Atomic Structure, Bonding and Periodicity*, section 10.3.2. **Electronegativity** is the power of an atom to withdraw electron density from a covalent bond. The data show that the electronegativity of the halogens decreases as the atomic number increases.

To explain this trend three important factors must be considered:

- the **atomic number**, which gives the nuclear charge. As this increases, the attraction for the bonding electron pair in a covalent bond might be expected to increase.

- the number of **electron shells**, which is indicated by the outer electron configuration of the atom. As this number increases, the shielding of the outer electrons from attraction by the nucleus increases. This results in the outer electrons being less strongly attracted.

- the **atomic radius** of the atom. The attraction between oppositely charged particles falls rapidly as the distance between them increases. As the radius of the atom increases, the outer electrons are further from the nucleus, which therefore attracts them less strongly.

The electronegativity of an element depends on a balance between these three factors. The changes in the values of electronegativity given for the halogens show that the increase in shielding and the increase in atomic radius more than compensate for the increase in nuclear charge.

The large electronegativity value for fluorine means that the bond between an element and fluorine is likely to be more polar than the bond formed between the same element and the other halogens.

Trend in boiling point of the halogens

All halogens exist as diatomic molecules, X_2. The attraction between these molecules in the liquid state is due to weak intermolecular forces called van der Waals' forces (see *Atomic Structure, Bonding and Periodicity*, section 10.3.3). These are caused by temporary fluctuations in electron density within the molecules, resulting in temporary dipole attractions between the molecules.

The magnitude of van der Waals' attractive forces increases with the size of the molecules. This fact explains why, as both atomic and molecular radii of the halogens increase with increasing atomic number, the boiling points of the halogens also increase, as shown by the data in Table 5.

11.5.2 *Trends in chemical properties*

The oxidising power of the halogens

When any reagent is oxidised, electrons are taken from it. The electrons are accepted by the oxidising agent, which is itself reduced. The trend in oxidising power of the halogens is characterised by a decrease from strongly oxidising fluorine to barely oxidising iodine.

The reasons for the decreasing trend in oxidising power from fluorine to iodine down Group VII are quite complex, in that they involve an overall balance of energies in the process shown below:

$$\frac{1}{2}X_2 + e^- \longrightarrow X^-$$

It is helpful to consider this process as occurring in three stages:

- the strength of the X—X bond (breaking to form X atoms in the gaseous phase)

- the affinity of an X atom for an electron (forming an X^- ion in the gaseous phase)

- the energy released when the X^- ion goes into solution or into a crystal lattice.

These three features affect the halogens differently:

- the very strong oxidising ability of the fluorine molecule can be attributed partly to the weakness of the F—F bond

- the electron affinity does not vary greatly from one halogen atom to the next, so has little effect on the relative oxidising power

- the fluoride ion, being the smallest, has the most to gain from being hydrated or entering a crystal, whereas the iodide ion, being the largest, benefits much less.

Thus, the trend is a decrease in oxidising power from fluorine to iodine:

$$F_2 > Cl_2 > Br_2 > I_2$$

The relative oxidising power of chlorine, bromine and iodine can be determined experimentally in the laboratory by a series of displacement reactions. In these experiments, aqueous solutions of the three halogens are added separately to aqueous solutions containing the other two halide ions. The results are given in the tables on page 30.

The F—F bond is weak because **E** of repulsion between non-bonding electron pairs on the two atoms. The electron pairs are close together because the F atoms are small.

The hydration and **E** crystallisation energies decrease as the size of the ion increases; being largest for the small F^- ion and least for the big I^- ion.

Fluorine is the strongest of all **E** oxidising agents.

Sea water contains a low concentration of bromide ions. Bromine is extracted from it by treating the sea water with chlorine. The liberated bromine is expelled from the water using air, and is then concentrated in a series of separate stages.

Table 6
The reactions of $Cl_2(aq)$ with $Br^-(aq)$ and $I^-(aq)$

Iodine is almost insoluble in water, but in the presence of iodide ions it dissolves to form the complex ion $I_3^-(aq)$ which is brown.

Halide ion	Observations	Conclusion	Equation
$Br^-(aq)$	yellow/brown	Br_2 displaced	$2Br^- + Cl_2 \longrightarrow$ $2Cl^- + Br_2$
$I^-(aq)$	brown colour and/ or black precipitate	I_2 displaced	$2I^- + Cl_2 \longrightarrow$ $2Cl^- + I_2$

Table 7
The reactions of $Br_2(aq)$ with $Cl^-(aq)$ and $I^-(aq)$

Halide ion	Observations	Conclusion	Equation
$Cl^-(aq)$	no change	Cl_2 not displaced	no reaction
$I^-(aq)$	brown colour and/or black precipitate	I_2 displaced	$2I^- + Br_2 \longrightarrow$ $2Br^- + I_2$

Table 8
The reactions of $I_2(aq)$ with $Cl^-(aq)$ and $Br^-(aq)$

Halide ion	Observations	Conclusion	Equation
$Cl^-(aq)$	no change	Cl_2 not displaced	no reaction
$Br^-(aq)$	no change	Br_2 not displaced	no reaction

Fluorine is far too dangerous to be used other than in specially equipped laboratories by specially trained staff. However, experiments using fluorine have been carried out and show that fluorine will oxidise all other halide ions to the halogen.

These results confirm the order of oxidising power by showing that:

- chlorine will displace bromine and iodine
- bromine will displace iodine but not chlorine
- iodine will not displace either chlorine or bromine.

11.5.3 *Trends in properties of the halides*

Trends in the reducing ability of the halide ions
The ability of halogen molecules to behave as oxidising agents by accepting an additional electron to form a halide ion was considered in Section 11.5.2. When a halide ion behaves as a reducing agent, it loses that electron to the reagent it is reducing; this process is the reverse of that in which a halogen molecule acts as an oxidising agent. The trend in reducing power of the halide ions shows a decrease from the strongly reducing iodide ion to the non-reducing fluoride ion.

$$I^- > Br^- > Cl^- > F^-$$

The reactions of sodium halides with sulphuric acid
The trend in the reducing power of the halide ions is shown in the reaction of solid halide salts with concentrated sulphuric acid. The oxidation state of sulphur in sulphuric acid is $+6$. This can be reduced to $+4$, 0 or -2 depending on the reducing power of the halide ion. Experimental results are given in Table 9.

NaX	Observations	Products	Type of reaction
NaF	steamy fumes	HF	acid–base (F^- acting as a base)
NaCl	steamy fumes	HCl	acid–base (Cl^- acting as a base)
NaBr	steamy fumes	HBr	acid–base (Br^- acting as a base)
	colourless gas	SO_2	redox (reduction product of H_2SO_4)
	brown fumes	Br_2	redox (oxidation product of Br^-)
NaI	steamy fumes	HI	acid–base (I^- acting as a base)
	colourless gas	SO_2	redox (reduction product of H_2SO_4)
	yellow solid	S	redox (reduction product of H_2SO_4)
	smell of bad eggs	H_2S	redox (reduction product of H_2SO_4)
	black solid, purple fumes	I_2	redox (oxidation product of I^-)

Table 9
The reactions of concentrated sulphuric acid with solid sodium halides

These reactions can be demonstrated in a laboratory fume cupboard but full safety precautions must be taken. Deriving equations for the reactions which occur provides valuable revision of redox reactions. **E**

Hydrogen fluoride is an extremely dangerous gas and, in the presence of water, will even etch glass.

These results indicate that:

- iodide ions can reduce the sulphur in H_2SO_4 from oxidation state $+6$ to $+4$, as SO_2, then to 0, as the element sulphur, and finally to -2, as H_2S

- bromide ions can reduce the sulphur in H_2SO_4 from oxidation state $+6$ to $+4$, as SO_2

- fluoride and chloride cannot reduce the sulphur in H_2SO_4 under these conditions.

Using silver nitrate and ammonia solution as a test for chloride, bromide and iodide ions in solution

Silver fluoride is soluble in water but silver chloride, silver bromide and silver iodide are all insoluble. Silver chloride, bromide and iodide are precipitated when an aqueous solution containing the appropriate halide ion is treated with an aqueous solution of silver nitrate. The colours of the three silver salts formed with chloride, bromide and iodide ions, and their different solubilities in aqueous ammonia, can be used as a test for the presence of the halide. These results are summarised in Table 10.

N.B. Adding dilute nitric acid to the solution under test before the addition of silver nitrate solution prevents the formation of other insoluble compounds, such as Ag_2CO_3.

Halide ion	Precipitate	Colour	Solubility of precipitate in ammonia solution
F^-	none	–	–
Cl^-	AgCl	white	soluble in dilute $NH_3(aq)$
Br^-	AgBr	cream	sparingly soluble in dilute $NH_3(aq)$, soluble in concentrated $NH_3(aq)$
I^-	AgI	yellow	insoluble in concentrated $NH_3(aq)$

Table 10
Testing for halide ions using $AgNO_3(aq)$ and $NH_3(aq)$

These results show that the solubility of the silver halides in ammonia solution decreases in the following order:

AgF > AgCl > AgBr > AgI

11.5.4 *Uses of chlorine and estimation of chlorate(I)*

The products obtained when chlorine reacts with water depend on the conditions used. Under normal laboratory conditions a very pale green solution is formed, showing the presence of the element chlorine, and an equilibrium is established as shown below:

$$Cl_2 + H_2O \rightleftharpoons HCl + HClO$$

This reaction is an example of a **disproportionation** reaction in which one species, in this case chlorine, is simultaneously both oxidised and reduced.

Oxidation state of chlorine: 0   −1   +1

$$Cl_2 + H_2O \rightleftharpoons HCl + HClO$$

If universal indicator is added to a solution of chlorine water, it first turns red since both the reaction products are acids, i.e. hydrochloric acid, HCl, which is a strong acid, and chloric(I) acid, HClO, which is a weak acid. The red colour then disappears and a colourless solution is left, because chloric(I) acid is a very effective bleach.

If chlorine is bubbled through water in the presence of bright sunlight, or the green solution of chlorine water is left in bright sunlight, a colourless gas is produced and the green colour, due to chlorine, fades. Tests show that the colourless gas evolved is oxygen. Under these conditions chlorine oxidises water to oxygen and is itself reduced to chloride ions.

$$2Cl_2 + 2H_2O \longrightarrow 4H^+ + 4Cl^- + O_2$$

Because this reaction is rather slow, it is best to leave an inverted test tube containing chlorine water in sunlight for several days, after which sufficient oxygen will have been produced to give a positive test with a glowing splint.

Water treatment

Chlorine and chlorine compounds are used in water treatment. For many years, small quantities of chlorine have been added to drinking water and to swimming pools in order to kill bacteria. Recently, the use of chlorine to treat water in swimming pools has declined and been replaced by granular calcium hypochlorite, calcium chlorate(I), which is less hazardous.

The concentration of chlorine in drinking water is approximately 0.7 mg dm^{-3}. Higher concentrations are used in swimming pools. Great care is taken to ensure that the correct amounts of chlorine are used because it is very toxic.

Reaction of chlorine with cold dilute aqueous sodium hydroxide

When chlorine reacts with cold water, an equilibrium is established between the reactants and the two acidic products. If water is replaced by cold dilute sodium hydroxide, the effect is to displace the equilibrium as the hydroxide reacts with the acids produced. The ionic equation for the reaction is given below:

$$Cl_2 + 2OH^- \longrightarrow Cl^- + ClO^- + H_2O$$

This reaction is of great commercial importance because the mixture of sodium chloride and sodium chlorate(I) is used as a bleach.

Sodium chlorate(I) is a powerful oxidising agent and, in acidified aqueous solution, will oxidise iodide ions quantitatively to iodine. The overall equation for this reaction can be deduced from two half-equations:

$$ClO^- + 2H^+ + 2e^- \longrightarrow Cl^- + H_2O$$

$$2I^- \longrightarrow I_2 + 2e^-$$

$$ClO^- + 2I^- + 2H^+ \longrightarrow Cl^- + H_2O + I_2$$

In a typical determination, sodium chlorate(I) is added to aqueous potassium iodide which has been acidified with dilute hydrochloric acid. The iodine liberated is determined by titration against a standard solution of sodium thiosulphate, $Na_2S_2O_3$. The ionic equation for this reaction is shown below:

$$I_2 + 2S_2O_3^{2-} \longrightarrow 2I^- + S_4O_6^{2-}$$

The characteristic brown colour due to the presence of iodine is not sufficiently intense to indicate the end-point. Starch, which forms a dark blue complex with iodine, is added near the end of the titration once the iodine colour has faded to a pale straw colour. The blue colour disappears at the end-point, leaving a colourless solution containing colloidal starch.

This analysis can be used in the laboratory to investigate the 'value for money' of a range of commercial bleaches. When commercial bleach containing sodium chlorate(I), NaClO, and sodium chloride is acidified with hydrochloric acid, chlorine is liberated. If an excess of potassium iodide is added first, acidification the chlorine will quantitatively displace iodine. This, being much less volatile than chlorine, does not escape as a gas. The amount of iodine liberated can be determined by titration against a standard solution of sodium thiosulphate.

Example

25.0 cm³ of a commercial bleach were transferred to a graduated flask and made up to 250 cm³ with distilled water. A 25.0 cm³ portion of this solution was added to a conical flask containing an excess of potassium iodide solution, and acidified with dilute hydrochloric acid. The liberated iodine was found to react with 18.7 cm³ of 0.100 M $Na_2S_2O_3$ solution. Calculate the mass of NaClO in a 750 cm³ bottle of this bleach.

Answer

Knowing the concentration and volume of the $Na_2S_2O_3$ enables the number of moles of $S_2O_3^{2-}$ used to be calculated.

$$\text{Moles } S_2O_3^{2-} = \frac{\text{molarity} \times \text{volume used}}{1000} = \frac{0.100 \times 18.7}{1000}$$

$$= 1.87 \times 10^{-3}$$

The reactions are:

$$ClO^- + 2I^- + 2H^+ \longrightarrow Cl^- + H_2O + I_2$$
$$2S_2O_3^{2-} + I_2 \longrightarrow 2I^- + S_4O_6^{2-}$$

The first equation indicates that:

1 mol of ClO^- forms 1 mol of I_2

The second equation indicates that:

1 mol of I_2 reacts with 2 mol of $S_2O_3^{2-}$

Since the number of moles of ClO^- is half the number of moles of $S_2O_3^{2-}$, 25.0 cm³ of diluted bleach contain $\frac{1}{2}(1.87 \times 10^{-3}) = 9.35 \times 10^{-4}$ mol NaClO.

Hence, 250 cm³ of diluted solution contain 9.35×10^{-3} mol NaClO. Since this solution was prepared from 25.0 cm³ of the original bleach solution, this is also the number of moles of NaClO in this volume of bleach. Since 25.0 cm³ bleach contain 9.35×10^{-3} mol NaClO, 750 cm³ contain

$$9.35 \times 10^{-3} \times \frac{750}{25} = 0.2805 \text{ mol}$$

Mass NaClO = number of moles $\times M_r = 0.2805 \times 74.5 = 20.9$ g

11.6 Extraction of metals

The production of metals is an important part of the chemical manufacturing industry. The way in which metals are produced from natural resources involves an understanding of the social and economic aspects of the processes, as well as an appreciation of the underlying chemistry.

Occurrence

One of the jobs of a chemist is to transform the lumps of rock all around us into useful materials. Metals are important materials – they are used for their strength, their ductility, and their thermal and electrical conductivities. Of the possible engineering metals, only aluminium and iron are very abundant, with titanium being the next most common. These three metals are also widely distributed, so they are the metals we might expect to be used whenever possible.

Aluminium is widespread in clay (an aluminosilicate), but a higher concentration of aluminium is found in bauxite, an ore which contains 50–70% of aluminium oxide. Nickel is found in concentrated form in a few places, for example in pentlandite at Sudbury, Ontario.

Some metals in everyday use, e.g. copper and nickel, are relatively rare in the Earth's crust. Fortunately, they sometimes occur in high grade ores in a few specific locations. To be economically viable, ores should have a high concentration of the desired metal and be free of impurities. It should be noted that ores of expensive commercial metals, such as copper, may have a low concentration of the metal whereas ores of lower value metals, such as iron, must contain very high concentrations of the metal to be economically viable.

Oxide and sulphide ores

Metals usually occur in combination with oxygen or sulphur. The process of extraction is therefore one of reduction – often of oxides. Because these ores are very stable compounds, energy usually must be put in to reduce them to the metal, i.e. extraction reactions are endothermic (unless a very powerful reducing agent such as aluminium metal is used).

Sulphide ores are not usually reduced directly to the metals. They are first roasted in air to produce oxides; this liberates sulphur dioxide and so gives rise to the potential pollution hazard of acid rain. Sulphur dioxide dissolves in water in the clouds to form sulphurous acid according to the equation:

$$SO_2(g) + H_2O(l) \longrightarrow H_2SO_3(aq)$$

Some of the sulphur dioxide is oxidised to sulphur trioxide:

$$2SO_2(g) + O_2(g) \longrightarrow 2SO_3(g)$$

This gas dissolves in water to form sulphuric acid:

$$SO_3(g) + H_2O(l) \longrightarrow H_2SO_4(aq)$$

These acids can then fall as acid rain, damaging plants and polluting lakes.

Oxide ores can often be reduced directly to the metal. However, the use of fossil fuels, such as coke from coal, may give rise to environmental problems through emission of carbon dioxide, a greenhouse gas.

Extraction

There are many ways by which metal oxides can be reduced to metals. The method actually chosen on an industrial scale for each metal depends upon several factors, including:

- the cost of the reducing agent
- the cost of energy for the process
- the required purity of the metal.

Clearly the overall cost of an extraction process is of prime importance. A process which requires less energy (e.g. operates at a lower working temperature) usually has an economic advantage over a process which requires greater energy input. A reducing agent that is naturally available, such as carbon (e.g. from coal) or hydrogen (e.g. from natural gas) is cheaper than one, such as aluminium, which has to be prepared by a separate (and often costly) process.

> Usually, carbon is cheaper than hydrogen and easier to store. Carbon is available as coke but hydrogen has to be prepared from methane or water.

The costs of the reductant and the energy requirements must be weighed against each other. If a metal is required with very high purity (and in small quantity) for a particular use, then a relatively expensive process may be used, the extra expense being needed to satisfy the demand for high purity and justified by the market value of the product. If high purity is not essential, then the cheapest method of producing saleable metal is used.

11.6.1 *Reduction of metal oxides with carbon*

Carbon is a cheap and plentiful reducing agent – it occurs in coal (which when heated in the absence of air gives coke, a solid with a very high carbon content) and in charcoal (which is obtained from wood, a renewable resource). All metal oxides can, in theory, be reduced by carbon if the temperature is high enough. In practice, temperatures over 2000 °C are impractical and uneconomic. The most important example of carbon reduction is the manufacture of iron from high quality haematite, Fe_2O_3.

> Because the reaction: **E**
> $$MO(s) + C(s) \rightarrow M(l) + CO(g)$$
> proceeds from left to right with the production of a gas, the entropy change, ΔS, for the reaction is positive. Using the relationship,
> $$\Delta G = \Delta H - T\Delta S$$
> we can see there will always be a temperature T at which $T\Delta S$ outweighs ΔH and at which ΔG becomes negative and the reaction becomes feasible.

Iron

Iron oxides are reduced by coke in a blast furnace. This is a continuous process in which iron(III) oxide, Fe_2O_3, coke and limestone are fed in at the top of the furnace and hot air is blown in near the bottom. Molten iron collects at the bottom of the furnace and is run off. The product is very impure iron which typically contains some 4% of carbon, as well as manganese, silicon, phosphorus and sulphur in amounts which depend upon the operating conditions and ore used.

Initially, coke reacts with the hot air blast in a strongly exothermic reaction:

> For an explanation of entropy see *Thermodynamics and Further Inorganic Chemistry*, section 14.1.2.

$$C(s) + O_2(g) \longrightarrow 2CO_2(g)$$

This reaction produces the heat needed for the reduction of the iron(III) oxide. The carbon dioxide formed reacts at high temperature with unreacted coke to form carbon monoxide:

$$CO_2(g) + C(s) \longrightarrow 2CO(g)$$

The carbon monoxide reduces most of the iron(III) oxide at around 1200 °C:

$$Fe_2O_3(s) + 3CO(g) \longrightarrow 2Fe(l) + 3CO_2(g)$$

In the hotter part of the furnace, coke also reacts directly with the iron oxide:

$$Fe_2O_3(s) + 3C(s) \longrightarrow 2Fe(l) + 3CO(g)$$

The above equation represents the overall reaction for the reduction.

The other product is a slag; this contains impurities from the ore combined with lime and is largely calcium silicate, $CaSiO_3$. Limestone is calcium carbonate; it decomposes in the heat of the furnace to form calcium oxide (lime) and carbon dioxide:

$$CaCO_3(s) \longrightarrow CaO(s) + CO_2(g)$$

Calcium oxide is a basic oxide and it combines with the acidic oxides in the furnace to form the slag. Sand and soil contain the acidic oxide silica, which reacts to give calcium silicate:

$$CaO(s) + SiO_2(s) \longrightarrow CaSiO_3(l)$$

Slag is used to make 'breeze blocks' for the construction industry.

Steel

Steels are iron–carbon alloys in which the carbon content is typically between 0.04% and 1.0%. The major process for converting iron into steel is the Basic Oxygen Steel-making Process (BOS). First of all, the iron is desulphurised using magnesium or calcium. These metals react with sulphur impurities to form sulphides which float on the iron as a slag.

$$Mg + S \longrightarrow MgS$$

Oxygen is then blown onto the molten iron, and lime or calcium carbonate is added; the oxygen reacts with carbon, which is removed as carbon monoxide. The oxygen also reacts with other impurities in the iron (e.g. silicon, phosphorus and manganese) to form oxides that react with the CaO to produce slag which floats above the iron. Sulphur cannot be removed by oxygen treatment since iron is oxidised in preference.

Not all of the carbon is removed, however. Pure iron is too soft for structural use. A small amount of carbon makes iron stronger, but too much carbon makes it brittle. Steels containing 0.1% carbon (in mild steel) to as much as 1.0% carbon (in high-carbon steels) are normally produced by the BOS process.

Carbide formation

Many other metals are extracted by reduction of their oxides with carbon. However, some metals form metal carbides rather than the metal itself, so that this is not a practical method for extracting these metals in a pure state. Carbides, which are important engineering materials and potential catalysts, can be made by this route. When titanium(IV) oxide is heated with carbon, the following reaction occurs:

$$TiO_2 + 3C \longrightarrow TiC + 2CO$$

Oxides of vanadium, tungsten and molybdenum also give carbides when they are heated with carbon.

Tungsten carbide is very hard.

Vanadium is prepared by silicon reduction of V_2O_5, and tungsten and molybdenum by reduction of their oxides with hydrogen.

11.6.2 *Reduction of metal oxides by the electrolysis of melts*

When the metal oxide is extremely stable, then electrolytic methods may be used. The manufacture of aluminium is carried out by the electrolysis of purified bauxite, Al_2O_3, which is dissolved in molten cryolite, Na_3AlF_6. The melting point of aluminium oxide is over 2000 °C, but by dissolving the oxide in molten cryolite, the temperature of the melt is reduced to about 970 °C. The electrodes are made of carbon and the reactions at the electrodes are:

at the cathode $Al^{3+} + 3e^- \longrightarrow Al$

at the anode $2O^{2-} \longrightarrow O_2 + 4e^-$

Some of the oxygen evolved reacts with the anodes at the high temperature:

$2C + O_2 \longrightarrow 2CO$

and

$C + O_2 \longrightarrow CO_2$

The process consumes large amounts of electricity (because electricity is needed to melt the cryolite and to decompose the Al_2O_3) and is only economic where electricity is relatively inexpensive. The process is continuous but regular additions of aluminium oxide are needed, and the carbon electrodes need replacing as they are consumed. There is a potential environmental problem through waste cryolite causing fluoride pollution.

11.6.3 *Reduction of metal halides by active metals*

When the purity of the metal is of prime concern and contamination from carbon or oxygen cannot be tolerated, then reduction of a metal halide with a reactive metal becomes a desirable route despite the cost.

Titanium is very abundant in the Earth's crust. It is also a very desirable engineering metal, having a low density, high strength and high resistance to corrosion. However, unlike iron – where small amounts of carbon and other elements can produce useful steels – traces of carbon, oxygen or nitrogen in titanium have undesirable effects (such as rendering the metal brittle). Titanium is therefore extracted from its chloride by reduction with an active metal.

The ore rutile, impure titanium(IV) oxide, is converted into titanium(IV) chloride using chlorine and coke at around 900 °C:

$TiO_2 + 2C + 2Cl_2 \longrightarrow TiCl_4 + 2CO$

Titanium(IV) chloride is a colourless liquid which fumes in moist air because of hydrolysis. It is purified from other chlorides (e.g. those of iron, silicon and chromium) by fractional distillation under argon or nitrogen. In the United Kingdom, the chloride is then reduced by sodium in the following exothermic reaction:

$TiCl_4 + 4Na \longrightarrow Ti + 4NaCl$

Elsewhere in the world, magnesium is used as the reducing agent in a similar reduction process. The magnesium chloride by-product is removed from the titanium by vacuum distillation at high temperature.

A batch process does not run continuously. Instead the titanium is made one batch at a time. This is more costly than running a continuous process like that in the blast furnace.

In the Electric Arc Process, a furnace is charged with scrap steel and an electric arc is struck between this scrap and some electrodes above the scrap. The heat generated melts the scrap. Lime is added and this combines with the impurities to form a slag which can be poured off.

The sodium is initially held at around 550 °C but the temperature rises to nearly 1000 °C during the reaction. An inert atmosphere of argon is used to prevent any contamination of the metal with oxygen or nitrogen. The sodium chloride by-product is washed out, leaving titanium as a granular powder.

This is a batch process. Costs are high because:

- chlorine and sodium have to be produced first
- high temperatures are involved in both stages of the production
- precautions have to be taken in handling $TiCl_4$, which reacts violently even with cold water
- an argon atmosphere has to be maintained to prevent oxidation.

Despite extensive searches, other methods for producing titanium have not yet been able to compete because of cost and purity considerations. So here is a metal with very desirable properties and a high natural abundance, yet its use is limited by the high cost of production.

11.6.4 *Economic factors and recycling*

The recycling of metals is carried out extensively. It would be environmentally friendly if all scrap was returned to the metal works and recycled. Unfortunately, because our scrap metal is often widely spread and many miles from metal-producing plants, the scrap must be collected and transported. This procedure creates an energy cost which must be calculated carefully and offset against the savings in extraction.

Iron (as steel) is the most extensively recycled metal; as much as 40% of the world's iron and steel production comes from scrap metal. The world reserves of iron ore are vast and commercial ores have a high iron content. So why recycle?

Firstly, the scrap iron contains a higher percentage of iron than a commercial ore. Many alloy steels are produced using only scrap metal in the Electric Arc Process. Also, in the BOS steel-making process, about 30% of scrap metal is added before adding the impure liquid iron.

Secondly, without recycling, a serious environmental problem would result from all the scrap metal that is discarded. About one-and-a-half million cars a year are discarded in the UK; about 75% of a car's mass is recycled, with only about 0.3% going to landfill. Steel cans are also recycled – because they are magnetic, they can easily be separated from other waste.

Thirdly, the extraction methods for both iron and aluminium produce carbon dioxide, a 'greenhouse gas'; re-melting does not produce this undesirable effect. Unlike the extraction of iron, the extraction of aluminium is very expensive because electrolysis consumes vast amounts of energy. The re-melting of aluminium cans, which are high-quality scrap, saves 95% of this extraction energy. It must be remembered, however, that there are considerable energy costs in collecting and separating the scrap aluminium, and these costs must be carefully balanced against the savings made in extraction.

AS 2 Sample module test

1 (a) The diagram below shows the Maxwell–Boltzmann energy distribution curves for molecules of a gas under two sets of conditions, **A** and **B**. The total area under curve **B** is the same as the total area under curve **A**.

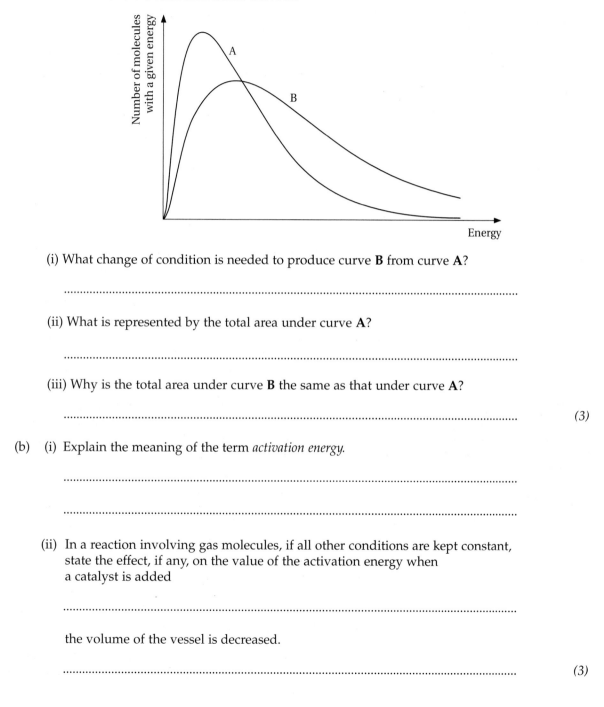

(i) What change of condition is needed to produce curve **B** from curve **A**?

....

(ii) What is represented by the total area under curve **A**?

....

(iii) Why is the total area under curve **B** the same as that under curve **A**?

....   *(3)*

(b) (i) Explain the meaning of the term *activation energy*.

....

....

(ii) In a reaction involving gas molecules, if all other conditions are kept constant, state the effect, if any, on the value of the activation energy when
a catalyst is added

....

the volume of the vessel is decreased.

....   *(3)*

(c) Explain why reactions between solids usually occur very slowly, if at all.

....

.... (2)

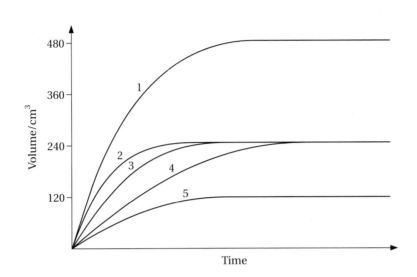

(d) The curves above, labelled 1 to 5, could be obtained by reacting an excess of zinc with different solutions of hydrochloric acid and measuring, at 25 °C, the volume of hydrogen produced.

The reaction of an excess of zinc with 100 cm^3 of 0.1 M hydrochloric acid at 25 °C gave curve 3. Which **one** of the curves, labelled 1 to 5, could have been produced by reacting excess zinc with each of the following? (Each curve may be used once, more than once or not at all.)

(i) 50 cm^3 of 0.2 M hydrochloric acid at 25 °C

....

(ii) 100 cm^3 of 0.1 M hydrochloric acid at 15 °C

....

(iii) 100 cm^3 of 0.2 M hydrochloric acid at 25 °C

....

(iv) 100 cm^3 of 0.1 M hydrochloric acid at 25 °C in the presence of a catalyst

.... (4)

(12)

2 (a) The enthalpy of combustion, $\Delta H_c^{\ominus}$, of carbon (graphite) is -394 kJ mol^{-1}.
What does the symbol $^{\ominus}$ indicate?

.... *(1)*

(b) In a simple experiment, 1.50 g of propane (C_3H_8), ($M_r = 44$) were completely burned in air. The heat evolved raised the temperature of 100 g of water by 65.0 K. Use these data to calculate the molar enthalpy of combustion of propane. (The specific heat capacity of water is 4.18 kJ K^{-1} kg^{-1}.)

....

....

....

.... *(4)*

(c) The combustion of propane occurs as shown in the equation below.

$$\begin{array}{c}
\text{H} \quad \text{H} \quad \text{H} \\
| \quad\;\; | \quad\;\; | \\
\text{H}-\text{C}-\text{C}-\text{C}-\text{H} \;+\; 5\text{O}{=}\text{O} \;\longrightarrow\; 3\text{O}{=}\text{C}{=}\text{O} \;+\; 4\text{H}-\text{O}-\text{H} \\
| \quad\;\; | \quad\;\; | \\
\text{H} \quad \text{H} \quad \text{H}
\end{array}$$

Use the following data to calculate the enthalpy change for the complete combustion of 1 mol of propane. Assume that all reactants and products are in the gaseous state.

Bond	C — C	C — H	O=O	C = O	H — O
Mean bond enthalpy/kJ mol^{-1}	348	416	496	743	463

....

....

....

.... *(4)*

(d) Give one reason why your answer to part (b) differs so much from the answer to part (c).

.... *(1)*

(10)

3   The equilibrium yield of a product in a gas-phase reaction varies with changes in temperature and pressure as shown below.

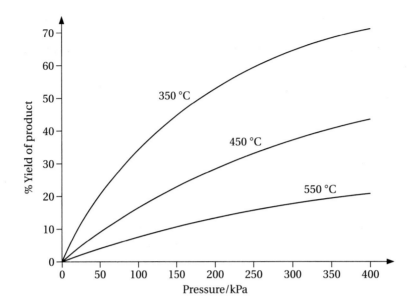

(a) Use the information given above to deduce whether the forward reaction involves an increase, a decrease or no change in the number of moles present. Explain your deduction.

Change in number of moles ....

Explanation ....

....

....   *(4)*

(b) Use the information given above to deduce whether the forward reaction is exothermic or endothermic. Explain your answer.

Forward reaction is ....

Explanation ....

....

....   *(3)*

(c) (i) Estimate the percentage yield of product which would be obtained at 350 °C and a pressure of 250 kPa.

....

(ii) State what effect, if any, a catalyst has on the position of the equilibrium. Explain your answer.

Effect on position ....

Explanation ....

....

.... (4)

(d) A 70% equilibrium yield of product is obtained at a temperature of 350 °C and a pressure of 400 kPa. Explain why an industrialist may choose to operate the plant at

(i) a temperature higher than 350 °C

....

(ii) a pressure lower than 400 kPa

.... (2)

(13)

4 (a) State the oxidation state of the named element in each of the following species.

(i) Nitrogen in nitrogen gas ....

(ii) Calcium in calcium hydroxide...

(iii) Copper in $[CuCl_4]^{2-}$ ....

(iv) Chromium in $Cr_2O_7^{2-}$ .... (4)

(b) When magnesium reacts with hydochloric acid, magnesium chloride and hydrogen are produced.

(i) Write an equation for this reaction.

....

(ii) State, in terms of electron transfer, what happens to a species when it is reduced.

....

(iii) Identify the reduction product of the reaction between magnesium and hydrochloric acid and write a half-equation for its formation.

Reduction product...

Equation .... (4)

(c) Sulphur dioxide, SO_2, can be oxidised to sulphate ions, SO_4^{2-}, in aqueous solution. Hydrogen ions are also produced. Write a half-equation for this reaction.

....

.... (2)

(10)

5   When 15.0 cm^3 of a solution of sodium chlorate(I) were treated with an excess of sodium iodide and the mixture then acidified, iodine was liberated. This iodine was found to react with 18.5 cm^3 of 0.150 M sodium thiosulphate.

(a) Use half-equations for the reduction of chlorate(I) ions and for the oxidation of iodide ions to deduce an overall equation for this reaction.

Half-equation for the reduction of chlorate(I) ions

....

Half-equation for the oxidation of iodide ions

....

Overall equation

.... (3)

(b) Write an equation for the reaction between iodine and thiosulphate ions, name an indicator used in this analysis and state the colour change at the end-point.

Equation ....

Indicator ....

Colour change at end-point .... (4)

(c) Calculate:

(i) the number of moles of sodium thiosulphate present in 18.5 cm^3 of 0.150 M solution

....

(ii) the number of moles of iodine liberated in the reaction

....

(iii) the number of moles of sodium chlorate(I) present in the original solution

....

(iv) the concentration of sodium chlorate(I) in mol dm^{-3}

.... *(4)*

(11)

6　Describe the tests you would carry out in order to distinguish between solid samples of sodium chloride and sodium iodide using

(a) silver nitrate solution and aqueous ammonia

(b) concentrated sulphuric acid.

For each test that you describe, you should state the observations that you would make and write an equation for the reaction which occurs. *(15)*

7 (a) State **three** methods for the extraction of metals from their purified ores and, by means of an equation or equations, give **one** example of a metal extracted by each method you choose. *(3)*

(b) Discuss the economics and environmental benefits of recycling iron and aluminium rather than extracting these metals from their ores. *(6)*

(15)

Module test answers

The symbol / indicates alternative acceptable answers.

1 (a) (i) increase temperature
    (ii) total number of molecules
    (iii) same number of molecules   3

(b) (i) minimum energy for reaction to occur
    (ii) *catalyst*   reduced
        volume   no effect   3

(c) particles cannot move about
    no collisions can occur   2

(d) (i) 2
    (ii) 4
    (iii) 1
    (iv) 2   4

2 (a) standard conditions (pressure 100 kPa, stated temperature)   1

(b) heat evolved $= m \times c \times \Delta T$

$$= \frac{100}{1000} \times 4.18 \times 65 = 27.17 \text{ kJ}$$

moles propane $= \dfrac{\text{mass}}{M_r} = \dfrac{1.5}{44} = 0.0341 \text{ mol}$

$$\Delta H = \frac{27.17}{0.0341} = 797 \text{ kJ mol}^{-1} \qquad 4$$

(c) bonds broken $= (2 \times$ C—C $+ 8 \times$ C—H $+ 5 \times$ O=O$)$
         $\Delta H = (2 \times 348) + (8 \times 416) + (5 \times 496)$
             $= 6504 \text{ kJ}$
    bonds formed $= (6 \times$ C=O $+ 8 \times$ H—O$)$
         $\Delta H = (6 \times 743) + (8 \times 463)$
             $= 8162 \text{ kJ}$
    enthalpy of combustion
         $= \Sigma \Delta H \text{ bonds broken} - \Sigma \Delta H \text{ bonds formed}$
         $= 6504 - 8162$
         $= -1658 \text{ kJ mol}^{-1}$ 4

(d) in part (b) heat is lost to the surroundings   1

3 (a) *Change in number of moles* decrease
    Explanation yield higher at higher pressure, hence equilibrium displaced to right or an application of Le Chatelier hence more volume/moles of gas on left-hand side of equation   4

(b) *Forward reaction* exothermic
    Explanation yield falls as temperature rises, hence equilibrium displaced to left-hand side by an increase in temperature or an application of Le Chatelier   3

(c) (i) 55–60%
    (ii) *Effect on position* none
        Explanation it increases the rate of forward and backward reactions equally   4

(d) (i) to increase the rate of the reaction/ reaction slow at 350 °C
    (ii) high pressure generation/equipment is expensive or cheaper at lower pressure, uneconomic to operate at higher pressure   2

4 (a) (i) zero
    (ii) +2
    (iii) +2
    (iv) +6   4

(b) (i) $\text{Mg} + 2\text{HCl} \longrightarrow \text{MgCl}_2 + \text{H}_2$
    (ii) species gains electrons
    (iii) *Reduction product* hydrogen
        Equation $2\text{H}^+ + 2\text{e}^- \longrightarrow \text{H}_2$ 4

(c) $\text{SO}_2 + 2\text{H}_2\text{O} \longrightarrow \text{SO}_4^{2-} + 4\text{H}^+ + 2\text{e}^-$ 2

5 (a) *Reduction of chlorate(I) ions*
     $\text{ClO}^- + 2\text{H}^+ + 2\text{e}^- \longrightarrow \text{Cl}^- + \text{H}_2\text{O}$
     Oxidation of iodide ions $2\text{I}^- \longrightarrow \text{I}_2 + 2\text{e}^-$
     Overall equation
     $\text{ClO}^- + 2\text{H}^+ + 2\text{I}^- \longrightarrow \text{Cl}^- + \text{H}_2\text{O} + \text{I}_2$ 3

(b) *Equation* $\text{I}_2 + 2\text{S}_2\text{O}_3^{2-} \longrightarrow 2\text{I}^- + \text{S}_4\text{O}_6^{2-}$
    Indicator starch
    Colour change blue $\longrightarrow$ colourless   4

(c) (i) $18.5 \times 0.150 \times 10^{-3} = 2.775 \times 10^{-3}$
    (ii) 1.387×10^{-3}
    (iii) 1.387×10^{-3}
    (iv) $0.0925 \text{ mol dm}^{-3}$ 4

6 (a) **NaCl** $\text{AgNO}_3 \longrightarrow$ white precipitate, dissolves in dilute $\text{NH}_3(\text{aq})$
          $\text{Ag}^+ + \text{Cl}^- \longrightarrow \text{AgCl}$
          $\text{AgCl} + 2\text{NH}_3 \longrightarrow \text{Ag(NH}_3)_2\text{Cl}$

     NaI $\text{AgNO}_3 \longrightarrow$ yellow precipitate, insoluble in concentrated $\text{NH}_3(\text{aq})$
         $\text{Ag}^+ + \text{I}^- \longrightarrow \text{AgI}$ (max 8)

(b) **NaCl** effervescence/colourless gas evolved, steamy fumes in air
         $\text{NaCl} + \text{H}_2\text{SO}_4 \longrightarrow \text{NaHSO}_4 + \text{HCl}$

     NaI effervescence, black solid/violet vapour, smell (pungent or bad eggs), yellow solid, steamy fumes

         $\text{NaI} + \text{H}_2\text{SO}_4 \longrightarrow \text{NaHSO}_4 + \text{HI}$
         $2\text{HI} + \text{H}_2\text{SO}_4 \longrightarrow \text{I}_2 + \text{SO}_2 + 2\text{H}_2\text{O}$ (max 7)

7 (a) MO + C + heat
Fe$_2$O$_3$ + 3C $\longrightarrow$ 2Fe + 3CO

Electrolysis of molten oxide
Al^{3+} + 3e$^-$ $\longrightarrow$ Al
2O^{2-} $\longrightarrow$ O$_2$ + 4e$^-$

Active metal reduction of a halide
TiCl$_4$ + 4Na $\longrightarrow$ Ti + 4NaCl

[Also MO + H$_2$, e.g. Mo; MO + Al, e.g. Cr; electrolysis
of molten chlorides, e.g. Na]   9

(b) **Fe** energy saving/high Fe content in recycled
material
Environmental benefits:
no dumps of waste, e.g. car bodies
no CO/CO$_2$ emission

Al saving electrical energy
no waste dumps of cans
no CO$_2$ from burning electrodes
no cryolite pollution possible   6

Collins Student Support Materials for AQA – ORDER FORM

This booklet covers one module from the AQA chemistry course at AS-level.
If you would like to order further copies from the series, please send a completed copy of this page to Collins by post or fax.

Title	ISBN	Price	Approval copy	Order quantity
1 Atomic Structure, Bonding and Periodicity	000327701 1	£3.99		
2 Foundation Physical and Inorganic Chemistry	000327702 X	£3.99		
3 Introduction to Organic Chemistry	000327703 8	£3.99		
4 Further Physical and Organic Chemistry	000327704 6	£3.99		
5 Thermodynamics and Further Inorganic Chemistry	000327705 4	£3.99		
TOTAL ORDER VALUE				

Details of other A-level titles in this series are available on our website:

www.**Collins**Education.com

Also available:

Collins Advanced Modular Sciences – Chemistry
comprehensive text books to support the new AQA specification

Title	ISBN	Price	Approval copy	Order quantity
Chemistry AS	000327753 4	£17.99		
Chemistry A2	000327754 2	£14.99		
TOTAL ORDER VALUE				

Please fill in your details and send your order to the address below:

Name	Tel:	0870 0100 442
Address	Post:	Collins Educational HarperCollins Publishers FREEPOST GW2446 GLASGOW G64 1BR